Polygonal Polyphony

A Composer's Guide To 3D Modeling Using Audiovisual Feedback For Enhanced Artistic Expression

Scarlett Beauregard Quentin,Ashworth

Table of Contents

1 Introduction to 3D Audio

Understanding the basics of sound is crucial. Think of sound waves, their frequencies determining pitch – high frequencies are high-pitched, low frequencies are low-pitched. Amplitude, the height of the wave, dictates volume – taller waves mean louder sounds. Then there's timbre, the unique quality of a sound, what makes a violin sound different from a trumpet, even at the same pitch and volume. Manipulating these three elements – frequency, amplitude, and timbre – is the foundation of effective sound design. Experiment with different sound sources; explore synthesis, sampling, and effects processing to shape your sounds. Remember, the goal is to craft sounds that perfectly complement your visuals. Consider the emotional impact each sound evokes – a high-pitched, shrill sound could suggest tension, while a low, rumbling sound might signify impending danger.

Now, let's move into spatial audio. Imagine sound isn't just coming from your speakers but from all around you. That's 3D audio. This creates a sense of immersion and realism. Techniques like binaural recording, which simulates how our ears perceive sound, are vital for achieving this. You can also use panning, which is moving a sound from one speaker to another, to place a sound in a specific direction. More sophisticated methods include using multiple speakers, or even headphones, to create a more complex soundscape. Delay and reverb effects are also your friends in spatial audio; they help to create the sense of distance and space. For instance, a sound with a long reverb tail placed subtly behind the main sound can suggest the sound source exists further away. The precise manipulation of these sound properties will define how realistic your spatial audio experience becomes.

Creating a believable soundscape is essential. Think about how sound behaves in real life. Sounds bounce off walls, creating echoes and reverberations. They also get absorbed by surfaces, affecting their volume and clarity. You can simulate these effects in your 3D audio using software plugins designed specifically for spatial audio processing. You'll want to experiment with room simulations and impulse responses, which essentially capture the acoustic characteristics of a specific space. By using the appropriate tools and sound effects to achieve a realistic and immersive sound, you can vastly enhance the overall quality of your work. Using these tools requires not only technical understanding but also creative exploration. Remember that each space has its unique acoustic features, influencing how sound propagates within it.

Positioning sound objects in a 3D environment involves more than just placing them randomly. Consider the listener's perspective. Imagine yourself standing in the middle of a scene. Where would you expect to hear the sounds coming from? Would they be directly in front of you, to your side, or behind you? Think about the size and shape of the environment and how that affects how sound travels. Consider the reflective surfaces of your space, the materials that absorb or diffuse the sound, and whether there are any obstacles in the way. It's about creating a believable relationship between the sound and the 3D space. The more you pay attention to these details, the more immersive and realistic your 3D audio will be. Accurate sound placement brings the scene alive and enhances the user's experience.

Beyond basic placement, consider how sound changes over distance. A sound source further away sounds quieter and often more muffled. You can simulate this attenuation in your 3D audio by adjusting the volume of the sound based on its distance from the listener. This seemingly small detail is often overlooked yet provides an extraordinary sense of realism to the audio experience. Experiment with different attenuation models until you achieve natural sound decay over distance, and remember that this also allows you to influence the overall atmosphere of your setting. Such manipulation of the audio helps to create an environment which is natural and credible.

Finally, remember the importance of testing. Different sound systems and headphones will reproduce your 3D audio differently. Use multiple testing methods and devices to ensure your sound is balanced and sounds good across a range of playback options. Get feedback from others to get a fresh perspective and catch errors you may have missed. Iteration is key. It's a continuous process of tweaking, refining, and experimenting until you achieve the exact experience you envision. Remember, a compelling 3D audio experience involves both technical expertise and an artist's understanding of the creative potentials that this methodology presents.

1.1 Sound Design Basics

Understanding the fundamentals of sound design is crucial for creating immersive and engaging 3D audio experiences. Think of it as painting with sound – you're not just creating noise; you're sculpting an auditory landscape. This involves mastering the manipulation of audio properties, such as amplitude, frequency, and timbre, to achieve specific expressive effects. Experiment with different sounds; try layering sounds, reversing them, or applying effects like reverb or delay to discover how they modify the perceived sonic space. Explore various sound libraries and synthesis techniques to find the perfect sounds for your projects. Remember, your goal isn't just to fill the space with sound, but to create a believable and emotionally resonant experience.

Effective sound design often relies on a careful blend of raw audio material and digital processing. Your initial recordings might be simple, perhaps the chirping of crickets or the rumble of a train, but using effects, you can drastically change their character and impact. Consider how equalization (EQ) can sculpt the sonic shape of a sound, enhancing certain frequencies and attenuating others to control the sound's tonal balance. Compression can control the dynamics, making quiet parts louder and loud parts quieter for a more even sonic texture. Reverb can simulate the acoustic space, immersing the sound in a virtual environment, while delay can add rhythmic echoes and depth. Mastering these tools allows you to mold sounds to fit seamlessly within your 3D environment.

The creation of impactful sound effects often involves manipulating existing sounds rather than using only pre-made samples. You may find yourself starting with a simple recording of a hammer blow, for instance. Through careful layering and processing—perhaps adding distortion for grit, or reverb to place it in a cavernous space—you can transform it into the sound of a collapsing building or a powerful magical strike. This approach encourages a deeper creative engagement, allowing you to personally sculpt every sonic detail rather than simply selecting pre-made options. Experimentation is key; don't hesitate to blend and distort sounds until they perfectly fit your vision.

Careful consideration of the sonic environment is essential for creating a truly compelling soundscape. Sounds don't exist in isolation; they interact with their surroundings. A simple bass note played in a small room will sound very different than the same note played in a large cathedral. This concept, known as acoustics, plays a critical role in creating realism and depth in your 3D audio designs. Experiment with using different reverbs and delays to emulate various acoustic spaces, creating the illusion of distance and proximity. Understanding how reflections and absorption affect sound is crucial; the material properties of a room – whether it's brick, wood, or concrete – all significantly alter the sound's overall character. By skillfully manipulating these elements, you can add significant texture and realism to your soundscapes.

To properly work with sound in the context of 3D modeling, it's essential to understand basic principles of audio mixing. Mixing involves carefully blending multiple sound sources to achieve a balanced and harmonious overall mix. This involves adjusting levels, panning sounds across the stereo field, and applying effects to individual tracks. Understanding how different sounds interact is key—for instance, how a high-pitched whistle might clash with a low-pitched drone. Remember, a good mix avoids muddiness, where different instruments become indistinct, and maintains a clear separation between individual sonic elements. A well-balanced mix ensures clarity and enhances the overall listening experience.

When constructing your soundscapes, think about the overall narrative you want to convey. What kind of atmosphere are you trying to create?

Is it a calm and peaceful setting, or a chaotic and exciting one? The sounds you select and how you arrange them will directly contribute to the overall mood and emotional impact of your work. This involves making careful choices, considering factors such as the tempo, rhythm, and intensity of the sound elements. Remember, sound can be just as powerful a storytelling tool as visuals. By using sound design effectively, you can deepen the emotional connection between the audience and the visuals, elevating the overall artistic expression.

1.2 Spatial Audio Concepts

Understanding spatial audio is crucial for creating immersive experiences. It's not just about making sounds louder or softer; it's about placing them precisely within a virtual 3D environment. This allows for a more realistic and engaging experience for the listener, enhancing the impact of your artistic vision. Imagine a soundscape where the listener can pinpoint the exact location of each instrument, a choir seemingly enveloping them, or a bird chirping subtly from a distant tree. This is the power of spatial audio.

Think of it like a three-dimensional canvas for your sound. Instead of a flat, two-dimensional soundscape, you're building depth and dimension. You can manipulate the perceived distance, height, and direction of sounds, creating intricate sonic environments. A simple example might be a helicopter approaching from behind the listener, the sound gradually growing louder and shifting from the rear to the front. This dynamic placement is what distinguishes spatial audio from traditional stereo or even surround sound.

Several techniques are used to achieve spatial audio effects. Binaural recording uses two microphones positioned like human ears to capture sounds with natural spatial cues. This provides a remarkably realistic listening experience, creating an accurate representation of the sound's location relative to the listener. Head-related transfer functions (HRTFs) are often employed in headphones and virtual reality applications; these are mathematical models that simulate how our ears and head shape the sound waves we perceive. This creates a realistic spatial sensation even when only using headphones. Ambisonics is another powerful approach, employing multiple speakers or channels to create a more encompassing 3D soundscape.

However, merely understanding the techniques isn't enough. Effective implementation requires careful consideration of the listener's perspective. The placement of virtual sound sources must be meticulously planned in relation to the 3D model. A misplaced sound source can disrupt immersion and create an unnatural listening experience. For example, a sound intended to appear to emanate from inside a virtual room, might instead sound as if it's originating from outside, compromising the illusion of reality.

Your choice of audio software and hardware can significantly impact the quality of your spatial audio. High-quality audio interfaces provide cleaner signals and more precise control over your sound sources, directly impacting the accuracy of your placement. Similarly, professional-grade headphones or speaker setups are essential for accurately monitoring and fine-tuning your work. Using subpar tools can lead to muddy or imprecise spatial representation, undermining the realism of your project. Remember, even the best spatial audio algorithms are only as good as the input signal and playback system used.

Beyond technical aspects, artistic considerations are paramount. The spatial positioning of sounds should serve the overall aesthetic of your work. Sounds should enhance the visual elements of your model, creating a harmonious audiovisual experience rather than a disjointed one. This might involve strategically placing sounds to guide the listener's attention to specific areas of the model, accentuating particular design elements, or simply enhancing the overall mood.

For beginners, a step-by-step approach is best. Start with simple experiments. Try placing a single sound source at various points in your 3D space and observe how the perceived location changes. Experiment with different HRTF settings and ambisonic configurations to explore the capabilities of your chosen tools. This hands-on approach will build your intuition and understanding of spatial audio's intricacies. Gradually, you will become more proficient in using spatial audio techniques to create fully immersive experiences, not just basic location cues.

As you become more familiar with these tools, experiment with the interaction between sound and visual elements. This is where the true power of this book's concepts become manifest. Consider how the listener's movement through a virtual space impacts their perception of the sound. Explore how dynamic sound placement can enhance interactive installations or create responsive animation within your 3D models. By connecting your soundscapes to your model's visuals, you are creating a more holistic and unified artistic statement. The goal is not to merely place sounds, but to weave them seamlessly into the very fabric of your visual creation.

2 3D Modeling Fundamentals

Understanding the fundamentals of 3D modeling is crucial for translating your sonic visions into tangible, three-dimensional forms. This chapter will equip you with the essential tools and techniques needed to build the visual foundations for your audio-reactive projects. Think of this as learning your instrument before composing a symphony.

We begin with primitive shapes. These are the building blocks of any 3D model—cubes, spheres, cylinders, cones, and planes. Most 3D modeling software provides these readily available, allowing you to quickly start constructing your designs. Experiment with their manipulation; resize, rotate, and reposition them to get a feel for the software's interface and the fundamental principles of three-dimensional space. This intuitive understanding will be invaluable as you progress to more complex forms. Remember, even the most intricate models begin with these simple primitives. Mastering them lays a solid foundation for more ambitious projects.

Next, we'll explore mesh manipulation. Once you have your initial shapes, the real creative work begins. Mesh manipulation involves adjusting the vertices, edges, and faces of your models to refine their form. Imagine sculpting clay; you add, subtract, and reshape until you achieve your desired form. Similarly, in 3D modeling, you'll use tools like extrude, bevel, and inset to modify existing geometry and create more detailed and complex structures. Practice these techniques extensively, as they are central to achieving precise control over your models' appearance. Consider starting with simple objects before tackling more complex forms. This gradual approach builds your confidence and understanding. Remember that patience and practice are key; it takes time to develop a refined skillset in mesh manipulation.

UV mapping is the process of "flattening" your three-dimensional model onto a two-dimensional surface. Why is this important? Because this flattened representation, your UV map, is how you apply textures—images, patterns, or colors—to your 3D model. Imagine wrapping a present; the wrapping paper is your texture, and the way it fits onto the present's surface is analogous to UV mapping. A well-constructed UV map ensures that your textures are displayed correctly on your model, preventing distortion or stretching. Learn to unwrap your models effectively, ensuring that your textures appear seamlessly and enhance the aesthetic quality of your final product. This seemingly technical process significantly impacts the visual appeal and realism of your work.

Now let's consider practical application. Imagine creating a 3D model of a speaker. You would start with basic primitives—perhaps a cylinder for the body and a cone for the woofer. Then, using mesh manipulation tools, you would refine the shape, adding details like buttons, ports, and a grill. Finally, you would create a UV map to apply a realistic wood texture to the cabinet and a metallic texture to the grill, transforming your simple primitives into a convincing, detailed 3D model. This hands-on approach reinforces the learning from this chapter. Remember to save your work frequently to avoid losing progress. This may seem obvious, but data loss is a common frustration for beginner modelers.

Throughout this chapter, we've focused on the core principles of 3D modeling, laying the groundwork for creating visually compelling models that will interact with your audio. Mastering these fundamental techniques will allow you to seamlessly integrate the visual and auditory aspects of your creative work. Remember that practice is paramount. Experiment with different techniques, try different software, and don't be afraid to make mistakes. Learning from those mistakes is an integral part of the creative process. The journey towards mastering 3D modeling is one of continuous learning and experimentation, a journey which will ultimately lead to the successful realization of your artistic visions. As you progress, you'll discover the true power of combining your musical sensibility with the expressive capabilities of 3D modeling.

Remember that each step in the process is crucial, and mastering the basics is your first step towards creating truly impressive and unique works. Your understanding of the fundamental techniques is the key to unlocking advanced techniques in later chapters. Consistent practice will reinforce these skills and build your intuitive sense of 3D space. Now, it's your turn to put these skills into practice. Start experimenting, and remember that continuous learning is the key to success in this field.

2.1 Primitive Shapes & Tools

Understanding the fundamental building blocks is crucial. Think of your 3D modeling software as a digital sculpting studio; you won't start by chiseling Michelangelo's David directly from a raw marble block. Instead, you'll begin with basic shapes – the primitives. These include cubes, spheres, cylinders, cones, and tori (doughnuts), each a simple geometric form that serves as a foundation for more complex creations. Mastering manipulation of these primitives forms the bedrock of your 3D modeling proficiency.

Your chosen software will likely present these primitives through a simple interface; a button, a menu item, or a hotkey will quickly generate them. Experiment with adjusting their dimensions – length, width, and height for boxes; radius and height for cylinders; radius for spheres. Play with these parameters; it's through this experimentation that intuitive understanding develops. Notice how subtle changes dramatically alter the visual impact. These adjustments are typically made through numerical input or by directly manipulating the shape using on-screen handles or gizmos provided by your software.

The tools associated with these primitives extend beyond mere creation. Most 3D modeling software offers a range of transformation tools. These allow you to move, rotate, and scale your primitive shapes. Mastering these foundational transformation tools is absolutely essential. Imagine you're building a house from Lego bricks—you need to be able to place each brick exactly where you want it, orient it correctly, and adjust its size if needed. It's the same principle in 3D modeling. Precise control over placement, rotation, and scaling is what transforms individual primitives into larger, more complex forms.

Beyond transformation, your software likely provides tools for adding subdivisions. Subdivision is a technique that increases the number of polygons in a mesh, allowing you to refine the shape and add detail. Imagine starting with a very low-resolution sphere, almost looking like a soccer ball. With subdivision, you can smooth out those angular facets, giving you a much more organic, smooth sphere. This process doesn't just affect spheres; you can apply subdivision to any primitive shape, opening up the potential for detailed forms. It's a non-destructive process, meaning the original shape remains intact, so experiment liberally; undoing your modifications is always an option.

Another important skill to develop early is the ability to combine primitive shapes. Imagine you want to model a simple chair. You might start with a cube for the seat, a cylinder for each leg, and a slightly curved rectangular prism for the backrest. By combining these basic shapes, manipulating their positions and scales, and potentially applying Boolean operations (explained later in the book), you can create a surprisingly realistic chair model. Practice combining basic shapes; it is through this creative process that you will begin to develop a visual language for your own designs.

It is important to remember the software-specific details. The exact names of tools and functions will vary depending on your choice of software (Blender, Maya, 3ds Max, Cinema 4D etc.). But the underlying concepts of primitives, transformation tools, and subdivision remain consistent. Familiarize yourself with your chosen software's interface, locate these crucial tools, and start experimenting. Create simple models and progressively increase their complexity. Don't be afraid to try different things—this is how you develop your own unique 3D modeling style. Learning to use these tools effectively is more than just understanding button placements; it's about developing a spatial understanding of the digital world you're creating.

Don't underestimate the power of simple shapes. Complex models are built from simple components. Mastering the fundamentals of primitives, transformations, and subdivisions allows you to generate an incredible variety of models; the capabilities extend far beyond basic shapes. As your proficiency grows, you will discover an intuitive understanding of how to generate complex structures using these fundamental tools. Your progress will be a testament to patience and focused practice, where even the simplest shape contains the potential for astonishing complexity. This is not just a technical exercise; it's the foundation of visual creativity in the digital realm. Remember, the more you practice, the more natural these tools will become.

2.2 Mesh Manipulation

Understanding mesh manipulation is crucial for refining your 3D models, transforming basic shapes into complex and expressive forms. It's like sculpting with digital clay, allowing you to add detail, smooth surfaces, and create intricate geometries impossible with primitive shapes alone. Mastering these techniques elevates your models from simple representations to nuanced artistic statements.

Think of a mesh as a collection of interconnected polygons, each a tiny facet contributing to the overall form. These polygons, usually triangles or quads, define the surface of your object. Manipulating the mesh involves altering the positions of these vertices, the points where polygons meet. This is fundamentally about editing the structure and shape of your digital object, not just its appearance via texture or colour.

Several tools are available for mesh manipulation, each offering different approaches to modify your model's geometry. Extrusion, for example, extends a selected face or edge along a specified axis, creating depth and volume. This is incredibly useful for generating simple details, like adding a rim to a cup or building up the layers of a complex structure. It's a cornerstone tool for beginners, allowing the creation of easily understandable forms and building confidence in this process.

Loop cutting, on the other hand, inserts new edges along the surface of your model, subdividing existing polygons and giving you finer control over curvature and detail. Imagine using a knife to slice through clay; you're creating new areas that can be shaped independently. This process creates new vertices that offer the opportunity to adjust areas of the model without changing its overall form.

Subdivision surface modeling offers a powerful way to refine meshes. It smooths out angular features by adding additional polygons between existing ones, creating smoother, more organic forms. Imagine a low-resolution image suddenly becoming sharp and high-resolution, yet keeping its fundamental structure. This technique is particularly useful for creating lifelike characters or complex architectural designs; the model starts with a simplified structure, allowing for quick editing and alterations, then smooths out into a beautiful, clean, well-defined mesh.

Edge loops, frequently used alongside loop cutting, allow for controlled shaping. You can create circular patterns that can be manipulated individually; imagine a series of concentric circles, each defining a level of shape control. By strategically placing these loops, you gain control over the flow of forms on your model's surface, resulting in more natural and realistic-looking bends and curves. This level of control provides greater depth and complexity compared to the basic methods.

Beveling edges softens sharp corners and transitions between surfaces. It adds a chamfer or fillet, rounding off edges and increasing the visual interest of your model. This is crucial for adding realistic detail to objects; consider the corners of a building where they might meet the wall, or the edges of a table which might be softened for safety or design purposes. This subtle change transforms a rigid, computer-generated model into something more organic and believable.

In many software packages, you'll find additional tools like scaling, rotating, and translating vertices directly, offering very granular control. These tools enable pinpoint accuracy when sculpting fine details, especially on highly complex models. You can literally move each point by small increments, creating the exact shape desired, without any limitation to the precision.

Beyond individual vertex manipulation, remember to consider the overall flow of your mesh. Look for areas that feel unnatural or abrupt. Use smoothing tools to blend transitions and enhance the visual appeal of your model. The overall shape of your mesh is what will define your piece in the end; understanding how the flow of polygons works is key in understanding the fundamentals of 3D modelling.

Experimentation is key. Try different techniques and tools to see how they affect your model. There's no single "right" way to manipulate a mesh, instead the process is iterative, requiring practice and exploration. You will find that the mesh manipulation techniques will allow you to achieve better results than you could with simpler techniques.

The final polish often involves a combination of techniques, refining the overall shape and adding fine details. Remember, mesh manipulation is not just about technical skill; it's about artistic expression, using these tools to bring your creative vision to life, one polygon at a time. The goal is to achieve a mesh that perfectly reflects the visual and sonic concepts you initially envisioned. By combining these different techniques, you can transform a simple mesh into an elaborate form.

2.3 UV Mapping Basics

UV mapping is the process of projecting a 2D image onto a 3D model's surface. Think of it like wrapping a present – you need to carefully flatten the wrapping paper (your texture) to fit the unique shape of the gift (your 3D model). Without proper UV mapping, your textures will appear distorted or stretched, ruining the visual appeal of your carefully crafted model. Mastering this crucial step is essential for creating professional-looking 3D assets.

This process involves unwrapping your 3D model's geometry into a 2D space, a process that requires careful planning to avoid excessive stretching or distortion. The goal is to create a UV map that maintains the proportions and details of your original model as accurately as possible. Improper UV mapping can lead to noticeable artifacts in your final render, making textures appear blurry, pixelated, or generally unrealistic. Many 3D modeling software packages provide automatic UV unwrapping tools. These tools use algorithms to flatten your model's geometry, often attempting to minimize distortion. However, relying solely on automatic unwrapping often produces suboptimal results. The algorithm might not understand the nuances of your model's design, leading to skewed proportions or unnatural seam placement. Understanding manual techniques and using the automatic tools as a starting point for refinement is a highly recommended approach.

Manual UV unwrapping, while more time-consuming, offers greater control and precision. You can carefully select seams, which determine how the 2D representation of your model's geometry is unfolded, ensuring that crucial details are maintained throughout the process. In practice, this means carefully planning the cuts in your model to result in a flat 2D image that's easily texurized and less distorted when it's re-mapped back onto the 3D surface.

Seams, the lines where your model's surface is conceptually cut to flatten it, are crucial to consider. Poor seam placement can dramatically impact the quality of your texture. Ideal seam placement often follows natural breaks in the model's geometry or areas where distortion is less noticeable. Experimentation and iterative refinement are vital to find the optimal seam placement for your specific model. A poorly chosen seam can stretch a texture in an unwanted way and make it look unnatural.

Once your model is unwrapped, you'll have a 2D representation of its surface, the UV map. This map dictates how your texture image is applied to the 3D model. The UV map is typically displayed as a square or rectangular area, and each point on this map corresponds to a point on your 3D model's surface. Understanding this correspondence is key to making the process work correctly. The UV map is often visualized in the UV editor window within your 3D software, and its boundaries can be adjusted for fine-tuning the texture projection.

Next, you import or create your texture image. This image should be carefully prepared in a 2D image editor to match the dimensions and aspect ratio of your UV map. Then, you 'bake' the textures, meaning you have the 3D model's geometry data and the UV map to determine how the texture is applied to each part of the model. This step ensures a seamless and consistent application of the texture to your 3D model. This is a crucial process that will take your model from basic shapes to a finished product.

Consider the resolution of your texture image. A higher resolution produces a sharper and more detailed texture, but it increases the file size of your final model. Choosing the right resolution depends on your project's requirements and the intended level of detail. It's important to strike a balance between quality and efficiency to avoid excessively large file sizes impacting performance.

Finally, rendering the model with the applied texture allows you to see the results of your UV mapping work. If you encounter distortions, you may need to adjust the UV map or your texture image to correct the problem. This iterative process, refining both the UV map and the texture to achieve the best possible outcome, is fundamental to mastering this essential 3D modeling skill. Don't be afraid to experiment and iterate until you are satisfied with the final product, making sure that the texture appears consistently and correctly across the model.

3 Visualizing Sound

Imagine sound as a sculptor's clay, malleable and responsive. This chapter focuses on transforming the intangible essence of sound into tangible 3D forms, allowing you to see the music. We'll explore several methods for translating the invisible vibrations of audio into visually stunning and conceptually rich models, going beyond simple waveform representations.

Frequency, the pitch of a sound, provides a rich source of visual data. Think of a simple sine wave, its smooth curve easily translated into a spiraling column whose height varies with the frequency. Higher pitches could be represented by taller, more tightly wound spirals, while lower tones might create broader, more gently sloping forms. Consider the potential of visualizing a complex chord, each individual frequency forming its own unique spiral, intertwining and interacting to create a visually captivating representation of the harmonic relationships within the sound. Imagine a complex piece of music represented not as a static score, but as a vibrant, evolving sculpture.

Amplitude, the loudness, offers another dimension for visual expression. This could be represented by the thickness or intensity of a line, a thicker line representing a louder sound, for instance. Imagine mapping this to a 3D model. A simple sphere could pulse and expand with the amplitude of the bass line, while intricate, smaller spheres could respond to higher frequency elements. This dynamic interplay provides a visual representation of the energy of the music, creating an organic visual equivalent to the feeling of the audio.

Spectral analysis allows you to visualize the frequency content of a sound across time. A spectrogram, a common tool in audio engineering, displays frequency on the vertical axis and time on the horizontal axis, with the intensity of each frequency at any given time represented by color or brightness. We can directly translate the spectrogram's data into 3D space. Visualize a landscape that evolves over time, with peaks and valleys representing dominant frequencies at different moments. This 3D spectrogram isn't merely a static representation; rather, it's a dynamic sculpture, an ever-changing terrain reflecting the musical journey.

Real-time feedback loops are crucial for creating dynamic 3D models that react instantly to the audio input. Imagine a scene where the amplitude of a sound controls the growth of a virtual plant, making the leaves sway and expand with each pulse of the bass drum. This responsiveness transforms the static 3D model into a living, breathing entity, directly mirroring the ebb and flow of the audio. This concept becomes even more powerful with the inclusion of user interaction: the user's actions trigger changes both in the audio and in the visual display, forming a dynamic relationship between sounds and images.

Data-driven design takes this process a step further, allowing algorithms to generate complex 3D structures based directly on the audio data. Consider a program that analyzes a song's rhythmic patterns, using that data to create a cityscape of buildings whose heights and shapes correspond to the rhythmic complexity, perhaps even using different materials and textures to reflect melodic shifts. This opens up entirely new possibilities for creative expression, allowing for generative art where the music is not simply reflected but actively shapes the final visual outcome. This sophisticated technique transcends basic visualization, allowing you to create 3D worlds organically built from the essence of sound itself. The audio becomes the blueprint for the visual landscape.

The techniques outlined in this chapter empower you to create truly innovative, visually striking 3D models that go beyond simple representation; they provide profound visual interpretations of music. These approaches, combined with techniques learned in other chapters, will allow you to push the boundaries of audio-visual artistic expression. By visualizing sound, you are not merely presenting your music, you are revealing its very structure and emotional core, opening up a new avenue for both artistic creation and listener engagement. Through the careful and considered use of these methods, your creations will become unique and immersive experiences, transcending the limitations of traditional musical representation.

3.1 Frequency Visualization

Frequency visualization is crucial for bridging the gap between audio and visuals in your 3D projects. Think of it as giving sound a tangible form, a visual representation that lets you directly interact with its properties. Understanding how to visualize frequency is essential for creating truly dynamic and responsive experiences. We're not just talking about simple waveforms here; we're talking about translating complex sonic data into compelling 3D models and animations.

Effective frequency visualization relies on transforming abstract audio data into concrete visual elements. Consider a spectrum analyzer; its vertical bars change in height according to the amplitude of different frequency bands. In a 3D environment, you could map those bars to the height of objects, the radius of spheres, or the intensity of emitted light. The possibilities extend beyond simple bar graphs, however. Imagine using frequency data to modulate the color of a textured surface, causing it to ripple and shift in response to specific sonic frequencies. This allows for richer visual interpretations that capture the nuances of your sound. Consider, for example, how the lower frequencies might affect the overall scale of the model, while higher frequencies control fine details such as surface texture or animation speed.

For beginners, start with simple visualization techniques. Many 3D modeling software packages offer built-in tools or plugins that can analyze audio and generate visual feedback. Experiment with these tools to understand how different frequency ranges are represented visually. Pay close attention to the mapping process; how does the audio data translate into changes in your 3D model's geometry, color, or animation? A well-chosen mapping will enhance the visual experience, while a poorly-chosen one can obscure the artistic intent.

Once you've grasped the basics, explore more advanced techniques. Consider using shaders within your 3D software to dynamically manipulate the visual aspects of your model based on frequency information. Shaders allow for highly customized visual effects. For example, you could create a shader that uses the frequency data to control particle systems, generating clouds of points that react to the audio in real-time. This allows for truly stunning visual displays. Remember that the key is to find a balance between technical accuracy and artistic expression. You might need to creatively interpret the frequency data to achieve a desired aesthetic outcome. This is where your artistic vision takes the lead.

Don't be afraid to experiment! Try different software combinations to achieve your desired effects. There's a wide variety of tools available, ranging from dedicated audio visualization plugins to general-purpose 3D modeling software with robust scripting capabilities. Experimentation is key, allowing you to discover what software and techniques best suit your creative style. Don't be discouraged if your initial attempts don't produce the perfect visual representation; refining your approach is an integral part of the design process. The visual representation of sound is a dynamic and evolving field; embrace experimentation to discover your unique approach.

Furthermore, keep in mind the context of your 3D environment. The effectiveness of your visualization will depend on how well it integrates with the overall design. A visually striking representation may be less effective if it clashes with the aesthetic of your 3D scene. Consider the color palette, lighting, and other visual elements when designing your frequency visualization. Your goal is to create a harmonious and immersive experience that integrates seamlessly. This ensures that the visual feedback works in sync with the design rather than distracting from it. A cohesive design enhances the overall impact of your project.

Finally, remember that frequency visualization is not just about technical proficiency. It's a creative process that requires artistic sensitivity and a deep understanding of both audio and visual design principles. As you become more experienced, you will develop your own unique style, finding ways to express your artistic vision through the dynamic interaction of sound and 3D visuals. Your own individual artistic style will eventually begin to shine through, creating a distinct and recognizable voice in your work. The visual interpretation of sonic data becomes a personalized tool, enhancing your creative expression. Ultimately, mastering frequency visualization will significantly elevate your 3D projects, creating truly immersive and captivating experiences.

3.2 Amplitude Visualization

Amplitude, the strength or intensity of a sound wave, offers a powerful visual language within 3D modeling. Unlike frequency, which dictates pitch, amplitude directly reflects volume. This chapter explores how to translate this dynamic quality into compelling visual elements that enhance your 3D models, creating a deeper connection between sound and form.

Consider a simple visualization: a sphere whose radius expands and contracts in direct proportion to the amplitude of the incoming audio. Loud sounds result in a larger sphere, softer sounds in a smaller one. This is a fundamental, yet effective way to map amplitude to visual size. Implementation is straightforward using many programming languages and 3D modeling software packages. Experiment with different scaling factors to find the optimal balance between subtle and dramatic changes.

Moving beyond basic scaling, you can map amplitude to a range of visual properties. The color of your model could fluctuate, shifting from a dark hue at low amplitude to bright at high amplitude, providing an immediate and intuitive sense of volume. Alternatively, you might consider adjusting the intensity of a light source shining on the model, creating a similar effect. These approaches allow for richer visual interpretations of your audio, and can be combined effectively for a more complex result.

Imagine an animated scene where several objects respond individually to different frequency ranges of a musical piece, creating a complex and engaging visual performance. Each object's scale or color might be dictated by the amplitude of that specific frequency band, while the positioning and movement are driven by other parameters, such as frequency or spectral content. In this way, you can create a truly layered and immersive experience, enhancing the overall aesthetic impact considerably.

Let's delve into more complex scenarios. You might represent amplitude through the density of particles. For low amplitude, the particle density is low; high amplitude creates a cloud of particles. This approach is particularly effective for visualizing ambient soundscapes, creating a dynamic visual representation of sonic textures. Think about implementing this within a procedural generation system; imagine a constantly evolving landscape that responds to the subtle changes in your soundtrack.

Another technique involves modifying the opacity of your 3D models. Low amplitude sounds would result in near-transparency, while high amplitude sounds would render the models fully opaque. This subtle approach can be surprisingly effective at conveying the dynamic nature of sound, particularly when coupled with other amplitude-based visual elements. Consider using this method on several overlapping layers of translucent geometries to create a complex visual interplay.

The key is to experiment and find methods that complement your artistic vision. Consider the specific nature of your sound and the overall aesthetic you aim to achieve. A chaotic soundscape might benefit from an equally chaotic visual representation; a delicate soundscape may require a more subtle visual approach. The relationship between sound and visuals should always be carefully considered, ensuring each enhances the other.

Your choice of software will also affect your implementation. Some tools provide direct interfaces for manipulating visual properties based on audio input, while others require more custom scripting. Regardless, the core concepts remain the same: map the amplitude data to visual changes in your model. The specifics of the implementation will depend on the software and your technical skills, but the underlying principle remains consistent. Learning to code will help to overcome constraints of ready-made solutions.

Remember that the visual representation of amplitude should always enhance, not distract from, your artistic vision. Overly complex or jarring visuals can negate the impact of even the most sophisticated sonic designs. Strive for clarity, elegance, and a seamless integration of the visual and auditory elements. Test your visualizations with a variety of audio sources to ensure your implementation is robust and versatile.

Finally, think beyond simple models. Consider how amplitude visualization could inform your design process. Perhaps you'll use the changing shape of a model to inspire new design ideas, generating new forms based on the dynamics of the audio input. This iterative process of design and visualization can lead to unforeseen creative results, pushing the boundaries of artistic expression within the realm of 3D modeling.

Experimentation is key. Let the audio guide your creation process, allowing the sonic landscape to shape the visual forms. Don't be afraid to break the rules; allow your own unique style to shine through. The possibilities are truly endless.

3.3 Spectral Analysis

Analyzing the spectral content of your audio is crucial for creating truly dynamic and responsive 3D models. Think of it as understanding the DNA of your sound; its constituent frequencies reveal its character and emotional impact, information you can directly translate into visual form. We're not just talking about basic waveforms here; we're delving into the intricate details of harmonic structure and timbre. This will inform your design choices in powerful ways.

Frequency analysis, at its core, involves breaking down a complex sound into its individual frequency components. Imagine a prism splitting white light into a rainbow – a similar process occurs when analyzing sound. Software such as Audacity, Reaper, or specialized plugins within your 3D modeling software (depending on your choice of platform) allow for visual representations of this frequency data. Pay close attention to the prominence of certain frequencies; are there strong bass frequencies that could be visualized as pulsating low-frequency movement in your model? Or perhaps high frequencies that could be represented by rapidly changing textures or colors? This is where your artistic interpretation comes into play.

Understanding the different visualization techniques is vital. Simple bar graphs or spectrogram displays offer different perspectives on frequency content. A bar graph provides a snapshot of amplitude at specific frequencies at a given moment, providing you with a clear understanding of the dominant frequencies. Spectrograms, on the other hand, illustrate how these frequencies change over time, revealing the dynamic evolution of the sound and allowing for the design of animations that precisely reflect this evolution. Experiment with various display methods, as each offers unique visual possibilities. You can use a frequency analysis to directly drive the animation, for instance, creating an animation directly driven by the amplitude of certain frequency bands.

Beyond frequency, consider the concept of spectral centroid. This metric represents the "center of gravity" of your sound's frequency spectrum. A high spectral centroid indicates a brighter sound, while a low one indicates a heavier, bass-rich sound. This single number offers a powerful tool for controlling global parameters in your 3D model. It could control the overall scale, the intensity of a light effect, or even the overall colour palette. This is especially important when working with polyphonic sounds that include multiple instruments, which will contain multiple spectral centroids to interpret.

Spectral flux, another critical aspect, measures the rate of change in the spectral content over time. This reflects the dynamic nature of your sound. A high spectral flux means a rapidly changing sound—think of a cymbal crash or a sudden shift in instrumentation—while a low spectral flux indicates a stable sound. This creates exciting opportunities for creating animations that capture the dynamism and intensity of the music. Imagine spectral flux driving the intensity of particles emitted from your 3D model, creating a stunning visual representation of the music's energy. This subtle shift in the energy of the music can create a significant impact on the viewer.

Don't limit yourself to pre-built visualization tools. Experiment with different data mapping techniques. You can map different spectral features to different aspects of your 3D model. For example, you might map the spectral centroid to the size of a sphere, the spectral flux to its rotation speed, and the overall energy to its color. This allows you to create complex interactions and unexpected visual outcomes. The key is to experiment and find combinations that work well for your chosen creative style.

Remember, this isn't just about technical accuracy; it's about creative interpretation. Use spectral analysis as a springboard for artistic expression. The data you obtain provides a foundation, but your aesthetic choices and design sensibilities ultimately dictate the final visual representation. By understanding the spectral nature of your sound, you can shape your 3D models with more precision and create an audiovisual experience that goes far beyond a simple visual representation of the music. Think about how you want the music to 'look' \item the feeling or the emotion you want your audience to feel - and design your visuals accordingly.

Developing an intuitive workflow is essential. You'll be constantly iterating between your audio editing software, your 3D modeling software, and potentially other tools for data manipulation. Finding an efficient way to exchange data and update visuals in real-time is key to productive creative work. It allows you to experiment freely and make quick adjustments. Don't be afraid to use scripting or other automation techniques to streamline your workflow, allowing you to focus on the artistic aspects of your work.

Finally, consider the limitations and possibilities of your software. Some software packages are better suited to specific tasks than others. Understand your tools' strengths and weaknesses and adapt your approach accordingly. This understanding will not only save you time, it will allow you to make informed choices about how to best utilize the spectral data you've gathered. You might want to even focus on developing a workflow between a few different software packages, choosing the best software for each individual task.

Ultimately, mastering spectral analysis will empower you to create truly innovative and immersive audiovisual experiences. It's a powerful tool that bridges the gap between sound and visual representation, opening up a vast realm of creative possibilities. Embrace experimentation, develop your own unique methods, and use this knowledge to amplify your artistic vision.

3.4 Real-time Feedback Loops

Imagine a shimmering sphere, its surface constantly morphing in response to the rhythm of a driving bassline. That's the power of real-time feedback loops in 3D modeling. This dynamic interaction between audio and visuals allows for the creation of breathtaking, responsive artworks that push the boundaries of traditional animation and sound design. It's not just about pre-programmed movements; it's about creating a living, breathing sculpture that reacts organically to sound.

We've explored the fundamentals of 3D modeling and sound design; now it's time to connect them. The key lies in understanding how to translate audio data into actionable parameters within your 3D modeling software. Think of it as building a bridge between two separate worlds—the world of sound waves and the world of polygons. This bridge requires careful planning and the right tools. Each step, from data acquisition to visual output, needs to be precise and efficient to keep the feedback loop smooth and responsive. This process demands a thorough understanding of programming fundamentals, allowing you to translate complex musical elements into simple, easily understood instructions.

Different software packages offer varying levels of support for real-time feedback. Some might require custom scripting, while others have built-in tools that streamline the process. Researching and mastering the capabilities of your chosen software is crucial. Don't be afraid to experiment. Trial and error are an integral part of the creative process. What works for one project might not work for another. Flexibility is key. Find the balance between utilizing pre-built tools to save time and leveraging the power of custom scripting for highly tailored solutions. The right balance is heavily dependent on your artistic vision.

Consider using frequency analysis to control the size or color of your objects. High frequencies might make your sphere expand, while low frequencies might cause it to pulsate slowly. Similarly, amplitude could influence the intensity of the visual effect. A loud sound might trigger a bright flash, while a quieter sound might produce a subtle shimmer. Explore the possibilities. Experiment with mapping different aspects of the audio spectrum to different parameters in your model. Consider spatial audio cues: let the position of a sound source influence the direction of an object's movement. This is where you bring true three-dimensional expression to your art.

This isn't simply about creating pretty visuals; it's about telling a story. Think about the narrative you want to convey with your artwork. How can your real-time feedback loop enhance this narrative? For example, a crescendo in the music might trigger a dramatic expansion of the model, while a quiet interlude could lead to a state of calm visual stillness. Consider how to layer various visual effects for a richer experience. Maybe adding subtle variations in texture based on the amplitude alongside the size changes.

Mastering real-time feedback loops isn't a destination; it's a journey of continuous learning and experimentation. It's a process of discovering new and inventive ways to translate sonic energy into visual spectacle. The ability to translate audio to visual information in real-time offers incredible freedom to the artist. Don't underestimate the learning curve. Be prepared for setbacks and troubleshooting. Patience is a virtue, but remember to celebrate small victories along the way. Your progress will build upon itself, and soon you will find yourself confident in crafting complex and unique audio-visual relationships. There is almost no limit to the creative potential this technique offers.

Remember the fundamentals: solid 3D modeling skills, a grasp of sound design principles, and a basic understanding of programming logic are essential. However, the most crucial ingredient is your own creativity. This chapter is a starting point, a guide to unlock the creative potential of real-time feedback. Now go forth and experiment, play, and create. The possibilities are as vast as your imagination. The future of audio-visual art is dynamic, interactive, and in your hands. Let your creative vision be the driving force behind your projects, shaping your unique voice and building a body of work that reflects your individuality and passion. Your next project should not simply be a demonstration of technique but a powerful expression of your artistic vision. The synergy of sound and form will allow you to explore new artistic territories and redefine what's possible. Don't be afraid to push boundaries and reinvent the way you express your creativity. Start small, build upon your successes, and watch your artistic skill grow exponentially.

3.5 Data-Driven Design

Imagine a shimmering sphere, its surface constantly rippling with light and color. This isn't magic; it's the result of data-driven design, translating the raw energy of sound into a breathtaking visual spectacle. We've already explored the basics of visualizing sound—frequency, amplitude, and spectral analysis—but this chapter dives deep into harnessing that information to shape your 3D models. Think of it as sculpting with sound itself, using its dynamic properties to inform every aspect of your creation.

Consider the possibilities: a towering structure whose height fluctuates with the intensity of the bass, or a swirling vortex whose speed corresponds to the tempo of the music. These aren't just aesthetic flourishes; they're opportunities to build a deeper, more meaningful relationship between the auditory and visual components of your work. By connecting the underlying data of a sound file directly to the parameters controlling your 3D model, you unlock a whole new realm of creative expression.

This process isn't about passively visualizing sound; it's about actively using it as a creative tool. You are no longer simply representing the audio data; you're using it as the driving force behind the aesthetic form. The algorithms you employ become your brushes, the sound data your palette. Experiment with different mapping techniques – how does frequency map to scale? What about amplitude to color intensity or texture detail? The choices you make will dramatically influence the final piece.

Let's consider a practical example: you might want your 3D model to react to the rhythmic elements of a song. You could map the beat to the animation of a specific element within the model—perhaps a pulsating light or a series of expanding and contracting shapes. The resulting effect is captivating, the visual rhythm mirroring the auditory pulse, enhancing the overall impact and creating a synchronized experience for the viewer.

However, this doesn't come without challenges. Choosing the right software and mastering the necessary programming skills are crucial. While some software offers pre-built tools for this purpose, you may find yourself needing to customize and extend these tools, potentially requiring some coding skills in languages like Python or C++. Don't feel discouraged by the technical hurdles. Start small, experimenting with simpler models and readily available tools before gradually integrating more complex aspects and data-sets.

The beauty of this approach lies in its iterative nature. Begin with a fundamental form, a simple cube or sphere, and gradually add complexity based on your audio input. Observe how your changes affect the final aesthetic and refine your approach accordingly. Don't be afraid to make mistakes; these are valuable learning opportunities. Each iteration brings you closer to a truly unique, dynamic, and responsive visual experience tightly linked to its auditory counterpart.

For instance, imagine using a granular synthesis algorithm to create a complex texture based on the frequency spectrum of an audio track. You can use these textures to construct intricate geometric patterns directly linked to the musical structure. Each note or chord will generate a distinct shape or texture, resulting in a fascinating dynamic system.

Another compelling approach involves exploring generative algorithms. These are essentially algorithms that can create entirely new data (shapes, textures, etc) based on the audio input. This opens up a whole new realm of creative possibilities, allowing you to create models that continually evolve and surprise the viewer. The key is to understand the relationship between the input audio and the desired output: a seamless integration of sound and visual form.

Remember that the effectiveness of your data-driven design hinges on the clarity of your artistic vision. Before diving into the technical aspects, clearly define the artistic statement you wish to make. What emotions or ideas do you want to convey? How can the interplay of sound and visuals best achieve this? A strong conceptual framework will serve as your compass throughout the design process.

Furthermore, effective data-driven design necessitates careful consideration of the audience. Think about how they will interact with your work. Will it be a passive experience, or will they have some control over the audio and visual feedback? Designing for the user is as crucial as designing the visual itself. User experience forms an integral part of the overall creative process.

Finally, don't be afraid to experiment with different visualization techniques. Explore the nuances of color palettes, particle systems, and geometric primitives. Each choice will significantly affect the mood and overall impact of the artwork. The possibilities are limitless; it's up to you to explore and discover the approach that best realizes your artistic vision. The real reward lies in the unique, expressive results you'll achieve by harnessing the power of sound to sculpt the very essence of your 3D model, creating a true marriage of audio and visual expression.

4 Audio-Reactive Animation

Animating 3D models in response to audio requires a solid grasp of animation principles. Think of it like composing music; you need rhythm, timing, and a sense of dynamics. Understanding these core principles—squash and stretch, anticipation, staging, follow-through, and overlapping action—will enable you to create believable and engaging movements. These principles, while seemingly simple, are fundamental to creating realistic and impactful animation that truly syncs with the audio. Mastering these will elevate your work beyond simple reactions to sophisticated, expressive sequences. Improper application can lead to stiff, unnatural movement, completely undermining the audio's emotional impact. Remember, the goal is not just to move objects, but to tell a visual story that complements the auditory narrative. This requires careful consideration of the entire animated sequence's pacing, emphasizing how the animation complements each musical phrase or sonic event.

Keyframing is your primary tool for synchronizing animation with audio. It's the process of setting specific poses or positions at key moments in your animation timeline. Software like Blender, Cinema 4D, or Maya provide tools to easily manipulate keyframes; you'll learn to adjust the timing and easing of these keyframes to perfectly match your audio. This is where the precision comes in. You need to carefully listen to your audio, identifying crucial moments—a drum beat, a vocal inflection, a change in melody—that need to be highlighted visually. Each of these moments provides an opportunity to create a synchronized visual beat. Don't just map keyframes directly to audio peaks; experiment with subtle delays and anticipations to improve timing and create more organic motion. The best animations often have a carefully considered interplay between direct correlation and subtle creative liberties.

Let's consider a practical example. Imagine you are animating a simple cube to react to a piece of music. You can tie the cube's scale to the amplitude of the audio. When the music is loud, the cube grows; when it's soft, it shrinks. This simple relationship between audio and animation can be achieved using various scripting approaches depending on your chosen software. Alternatively, you could make the cube rotate based on the frequency; higher frequencies translate into faster rotations. The possibilities are vast. Explore different visual parameters. Color changes can match energy levels, position shifts can highlight instrumental changes.

However, relying solely on direct audio-to-animation mapping can feel mechanical. The key is to blend technical precision with artistic intuition. Experiment with creating anticipation in your animations. Before a loud beat, make the cube slightly shrink, creating a sense of anticipation which mirrors the building tension in the music. Similarly, add follow-through to create a more natural and less robotic movement. These small adjustments often make the difference between static, lifeless animation and dynamic, expressive artwork. This delicate balance requires a deep understanding of both audio and visual storytelling.

Consider also different animation styles. Motion graphics offer a distinct approach, utilizing simplified, geometric shapes and bold movements. This style works well with electronic music or abstract soundscapes, creating a vibrant and energetic visual representation of audio. In contrast, more realistic or subtle animation might be better suited for classical music or ambient sounds, offering a softer response to audio cues. The most successful approach blends these styles creatively.

Your choice of software will impact your workflow, but the core principles remain the same. Consider the capabilities of various software packages—each software has its strengths and weaknesses. Some offer more robust scripting options for creating intricate audio-reactive animations. Experiment to discover the tools that best suit your creative style and the complexity of your project. Regardless of your choice, focus on clear organization of your project files, efficient keyframing techniques, and consistent referencing back to your audio source for successful synchronization. This streamlined workflow will allow you to focus on artistic nuances rather than technical obstacles. Experimentation and iteration are crucial.

Remember, the goal isn't to create a perfect, literal representation of the audio. The ideal outcome is an evocative visual interpretation that enhances the emotional impact of the music. This is where your artistic vision takes center stage. Use the audio as a starting point, a catalyst for your creative expression. Don't be afraid to experiment, break the rules, and develop your unique style in your audio-reactive animation practice. The path to mastery lies in consistent practice, experimentation, and a deep engagement with both the audio and the visual elements of your work. The combination of technical skill and artistic sensibility will result in compelling and truly unique animations.

4.1 Animation Principles

Squash and stretch. This fundamental principle, often overlooked, is the cornerstone of believable movement. Think of a bouncing ball; it doesn't just move from point A to point B. Instead, it compresses on impact, storing potential energy, then stretches as it rebounds, releasing that energy. This exaggeration of form gives the illusion of weight and elasticity, enhancing realism even in fantastical scenarios. Mastering squash and stretch isn't just about adding visual flair; it's about imbuing your animations with a tangible physics-based feel, lending credibility to your artistic choices. Successful application demands close observation of real-world movement, paying meticulous attention to how objects behave under different forces and conditions.

Anticipation is the setup. Before any significant action, a subtle hint of what's coming provides a satisfying sense of anticipation. A character about to jump, for instance, might crouch slightly, preparing the audience for the release of energy. It's this preparation, this subtle pre-action, that creates a more dynamic and visually satisfying effect, enhancing the impact of the primary action. This principle ensures the animation feels fluid, natural, and prevents the abruptness that can often ruin the overall flow. Without proper anticipation, even perfectly rendered movements will lack impact and organic grace. The art lies in finding the delicate balance; anticipating too much will become predictable, too little results in jerky and jarring action.

Staging is about clarity. Your animation, no matter how complex, should be immediately understandable. The viewer shouldn't have to work to comprehend the action. This means employing clever camera angles, carefully composing your shots, and utilizing visual cues to draw attention to the key elements of the scene. Staging isn't merely about positioning; it's about leading the eye, focusing attention where it's needed most, and preventing visual clutter from distracting from your main point. Think of it as directing the audience's attention to the important narrative moments, guiding their experience for maximum impact and clarity. It's about making sure that every frame tells a story.

Follow through and overlapping action prevent stiffness and add nuance. Think of a cat's tail—it doesn't stop moving the instant the cat does. This lingering movement, this overlapping action, adds life and realism to the animation. It's in these subtle details that a sense of weight, momentum, and organic motion comes to life. The principle of follow-through and overlapping action works equally well for inanimate objects. A pendulum swinging doesn't halt its momentum instantaneously, instead it gradually comes to rest. Mastering this technique adds layers of visual complexity, increasing your animation's verisimilitude, regardless of style.

Slow in and slow out—this is about creating natural acceleration and deceleration. Objects rarely move at a constant speed. They accelerate and decelerate, lending a feeling of ease and grace to the movement. Imagine a character walking; their pace shouldn't be uniform. The beginning and end of the motion should be slower, creating a more natural and engaging animation that is believable. This is easily accomplished through careful keyframing, paying close attention to the speed and timing of each frame. It's subtle, but the effects on the viewer are substantial; making the animation feel smoother and more organically driven, eliminating jerky or unnatural movement. The key is in subtle control, preventing jarring transitions.

Arcs are fundamental to natural movement. Rarely does anything move in a straight line. From the trajectory of a thrown ball to the swing of a pendulum, movement naturally follows curved paths, and understanding this improves the believability of your animations significantly. These gentle curves make for smoother motion, adding a fluidity that's often missing in animation that adheres too rigidly to straight lines. This principle, while simple, can dramatically impact the smoothness and realism of your 3D animations. Every element should exhibit some degree of natural curve, even in the most stylized of works.

Secondary action—this enhances the primary action, adding a layer of depth and interest. Think of a character walking and simultaneously swinging their arms; the arm movements are secondary actions that enrich and enhance the primary action of walking. It's about subtlety, creating layers of movement that organically complement the main action. These small details add a dynamic layer of complexity that elevates your animations. These secondary actions can create a stronger visual rhythm, adding to the scene's liveliness and impact. The secondary movements can tell stories on their own, enhancing the primary action while subtly supporting the core narrative.

Timing is critical. It determines the speed, rhythm, and overall feel of your animation. A fast motion communicates energy and excitement, whereas a slow motion conveys deliberation or suspense. It's about understanding the relationship between speed and emotion, and using that relationship to communicate the intended mood. Mastering timing is about understanding how pace relates to the narrative and character personality, shaping the overall experience. Precise timing elevates ordinary animation to a higher artistic plane, crafting a performance instead of mere movement.

Exaggeration emphasizes and heightens impact. It's about pushing the boundaries of reality to communicate emotions, add visual interest, and create an unforgettable impact. While realism is important, a touch of exaggeration can make your animations more memorable and emotionally resonant. It's about careful and judicious use, emphasizing important moments while enhancing the aesthetic appeal. Exaggeration can enhance a character's personality, making them immediately recognizable and memorable.

Solid drawing creates the illusion of three-dimensionality. Even in a digital environment, understanding the principles of form, weight, and perspective will elevate the quality of your work. Good drawing skills are transferable—even to 3D modelling. It's not just about visual accuracy; solid drawing is about portraying volume and weight in a convincing and visually engaging way. It's this understanding that helps you to craft realistic and immersive 3D worlds. The foundational elements of drawing translate directly to understanding volume and form in 3D space.

These twelve principles are not isolated techniques; they work together synergistically. By mastering them, you will elevate your animation from simple movement to captivating storytelling. Your animations will become more expressive, engaging, and believable, allowing you to effectively communicate your artistic vision. Remember, animation is about more than just moving objects; it's about crafting compelling narratives and emotionally resonant experiences.

4.2 Keyframing Techniques

Keyframing is the cornerstone of bringing your audio-reactive 3D models to life. It allows you to precisely control the movement of your objects, synchronizing them perfectly with the rhythm, dynamics, and even the harmonic content of your audio. Think of it as meticulously choreographing a dance where the music dictates each step. This precise control offers a level of detail and expressiveness unmatched by other animation techniques. Master it, and you unlock a whole new dimension in your creative potential.

Imagine a simple sphere reacting to a bass drum. You wouldn't want it to simply bounce randomly; you want a controlled, powerful pulse that mirrors the impact of the drum. Keyframing provides this control. You strategically place keyframes—snapshots of the sphere's position—at specific points in time. Between these keyframes, the software smoothly interpolates the movement, creating a fluid and natural animation synced with your audio. Experiment with different interpolation methods to fine-tune the sphere's response. Ease in and out of movements for a more organic feel, mimicking the way real-world objects accelerate and decelerate.

The power of keyframing extends beyond simple positional changes. You can keyframe almost any aspect of your model's properties: rotation, scale, even material properties like color or texture. For example, imagine visualizing a melody using a series of cubes that change color with the notes. Each note triggers a keyframe, altering the color of the corresponding cube. The result: a stunning visual representation of the musical progression. This opens up a vast array of possibilities for creative expression. Consider using keyframing to control the intensity of light sources, to subtly shift the position of a camera, or to manipulate transparency, all in synchronization with the audio. This intricate level of control allows you to create truly breathtaking visuals.

Working with audio requires understanding your waveforms. While the visual representation of the audio itself can inform your keyframing choices, you can also extract specific data points from the audio signal and use them to drive your keyframes. For instance, the amplitude of the audio can determine the scale of an object, increasing its size during loud passages and shrinking it during quiet sections. Similarly, you can map frequency data to control rotation speed, creating a visual representation of musical pitch. This detailed control allows for a nuanced and precise translation of sound into visual movement. The possibilities are nearly limitless, and you can get quite creative with how your project visualizes the audio.

Different software packages offer varying keyframing workflows. Some utilize a timeline-based approach, letting you place keyframes directly on a visual representation of your audio. Others might rely on more abstract methods, allowing you to manipulate parameters via curves or graphs. Familiarity with your chosen software's interface is crucial, enabling you to work efficiently and intuitively. It's crucial to develop an understanding of these various approaches, allowing you to choose the one that best suits your artistic vision and workflow. Experiment with various strategies and determine the most efficient one for your needs.

Don't limit yourself to simple, linear movements. Experiment with curves and more complex animations. Use easing functions—these allow for smoother transitions—to make your animations look more natural. Try combining multiple types of keyframing—position, rotation, scale—to create a rich and engaging visual experience. Even minor adjustments can significantly impact the overall impact and mood of your animation. Don't be afraid to iterate, refining and tweaking your keyframes until you achieve the exact visual effect you're after.

Remember, keyframing is an iterative process. It's rare to get it perfect on the first try. You'll likely need to adjust keyframes, experiment with different timings, and refine your animation multiple times to achieve the desired effect. Be patient, experiment, and embrace the iterative nature of this process; it is essential to develop a refined and visually compelling animation that accurately reflects the audio input. The final result will be worth the effort. The beauty lies in this process of refinement, this interplay between your artistic vision and the technical precision of your keyframing. The final product will reflect the time and effort you have put into it.

The most critical aspect of mastering keyframing is understanding your workflow. Develop a method that allows you to easily view your audio and visual elements simultaneously. Efficient organization is crucial, particularly when working with numerous keyframes. Use labeling and color-coding techniques to organize your keyframes and easily identify different elements in the animation. A well-organized workflow will allow you to focus on the creative aspects of the animation, rather than getting bogged down in technical details.

Finally, always keep your overall artistic vision in mind. Keyframing is not just about technical precision; it's about expressing your creative ideas through movement. The best animations are those that seamlessly blend technical skill with artistic vision, creating a visually compelling experience that enhances the emotional impact of your audio. Never underestimate the power of subtle movements or well-placed pauses. These can significantly impact your composition and audience experience. Your technical skill and artistic sensibilities must complement each other; only then can you create compelling animations. Remember, it's the intersection of meticulous technical control and profound artistic intent that yields exceptional results.

5 Advanced Modeling Techniques

Procedural modeling offers a powerful alternative to manual modeling, especially when dealing with complex or repetitive structures. Imagine needing to create a field of swaying grass blades—a painstaking task by hand. Procedural generation, however, allows you to define rules and parameters for the individual blades, their size, curvature, and animation, then let the computer generate thousands, each unique yet consistent with the overall aesthetic. This is achieved through scripting languages like Python within your chosen 3D modeling software, requiring you to learn the basic syntax and functions to manipulate geometry. Understanding nodes and how they influence each other is crucial for success, as is effective error handling and debugging your scripts. Experiment with different noise functions to introduce organic variation into your procedural textures, ensuring your generated elements look natural and avoid overly repetitive patterns. Efficient procedural modeling techniques save significant time and effort, resulting in more intricate and visually stunning results than would be possible with purely manual methods. Mastering procedural techniques enhances your overall efficiency and creative potential.

Sculpting, in contrast to the precise mathematics of procedural modeling, embraces a more intuitive and organic approach. Digital sculpting software provides tools similar to real-world sculpting materials, allowing you to shape 3D models using brushes, smoothing tools, and other instruments. This methodology is particularly well-suited for creating organic forms like characters, creatures, or natural landscapes, where precise geometry may not be as critical as the overall form and feel. The ability to easily add and remove material, refine details, and smoothly transition between areas of high and low detail is a huge advantage. This chapter will show how to implement dynamic topology sculpting and re-topology. Experiment with various brush settings, strengths, and sizes to achieve different effects; refine your models iteratively. This iterative process allows for continuous refinement and adjustment, enabling artists to achieve highly realistic or stylized results depending on their desired aesthetic. Remember the importance of proper base mesh preparation for optimal sculpting results.

Boolean operations represent a powerful yet surprisingly simple way to manipulate existing models. These operations—union, subtraction, and intersection—allow you to combine, carve away from, or find the overlapping volume of multiple 3D objects. Think of it like creating complex forms by assembling, removing, and modifying simpler building blocks. For example, you could use the subtraction operation to carve out a window from a wall, or the union operation to merge two separate parts into a complete structure. The possibilities are extensive, allowing for creative shortcuts and efficient generation of complex forms. Learning to anticipate and manage potential issues like non-manifold geometry that can occur after Boolean operations is crucial for a polished final result. Careful planning of your modeling approach is vital for seamless and error-free integration. Precise alignment and manipulation of objects prior to applying Boolean operations are paramount; otherwise, unwanted results may occur. By combining Boolean operations with other techniques discussed in this book, you can create remarkably intricate and detailed models.

5.1 Procedural Modeling

Procedural modeling offers a powerful alternative to manual 3D modeling, especially when dealing with complex, repetitive structures or organic forms. Instead of painstakingly creating each element individually, you leverage algorithms to generate geometry based on defined rules and parameters. Imagine creating a field of swaying grass blades, each subtly different yet consistent in its overall form, all without manually modeling thousands of individual blades. That's the power of procedural generation. This chapter dives into the core concepts and techniques you'll need to master this transformative method.

Understanding the fundamental principles of procedural generation involves grasping the concept of iterative processes. Think of it like a recipe: you start with basic ingredients (parameters like density, scale, and randomness), and through a series of steps (algorithms), you create a final product (the 3D model). These algorithms can range from simple mathematical formulas to more complex systems, often involving feedback loops to create intricate, evolving shapes. The key is learning how to define these rules effectively so the computer does the heavy lifting for you, allowing your creative vision to drive the process.

One of the most accessible methods of procedural modeling utilizes noise functions. These mathematical functions produce seemingly random yet structured patterns, ideal for generating textures and organic forms. Perlin noise, for instance, is widely used to create realistic terrain, clouds, or even the surface detail of a character's skin. By manipulating the parameters of the noise function—like frequency, amplitude, and octaves—you can control the level of detail and the overall character of the generated texture. Experimentation is key here; subtly shifting parameters can dramatically alter the final output, offering vast creative possibilities. Consider visualizing sound waves using this approach: high-frequency sounds might map to highly detailed, textured surfaces, while lower frequencies could translate to smoother, broader forms.

Beyond noise functions, you can employ fractal geometry to generate complex, self-similar structures. Fractals are mathematical sets that exhibit self-similarity across different scales. Think of a branching tree: each branch resembles the overall structure of the entire tree, but on a smaller scale. Algorithms can replicate this self-similarity, allowing the creation of incredibly detailed models with relatively simple code. This is especially useful for creating natural elements like trees, plants, or even the intricate branching of a river system. Your audio could dictate the branch density of a virtual tree, making for a dynamic, reactive artwork.

Another crucial aspect involves utilizing particle systems. These systems simulate the behavior of a large number of individual particles interacting with each other and their environment. Think of fireworks, smoke, or even a flock of birds. By setting parameters like particle velocity, lifespan, and interaction rules, you can generate complex shapes and animations that respond dynamically to your input. For instance, you could use a particle system to visualize the frequencies of sound, where higher frequencies generate more particles, creating denser clouds or visually exciting bursts of color. Consider the possibilities of a visual response to a musical crescendo – the particles could erupt in a dynamic display, then settle back down with the music.

However, successful procedural modeling isn't just about choosing the right algorithms; it's also about effectively managing parameters. This often involves experimenting and iteratively refining your approach. Start with simple models and gradually increase complexity. Understanding how adjustments to parameters affect the final output is crucial. Visualizing this process can be facilitated by using live-coding environments and integrating real-time feedback loops. Seeing the immediate results of parameter adjustments allows for intuitive and efficient design, making the process of iterative refinement both engaging and effective. Consider linking your parameters to specific audio properties – a musical key change could trigger a change in procedural generation, making for a truly dynamic and responsive system.

Finally, bear in mind that procedural modeling isn't about replacing traditional modeling entirely; rather, it's about augmenting it. It's a powerful tool for creating complex base meshes or intricate details, but you might still need to utilize other techniques—like sculpting or Boolean operations—to refine and polish the final result. Integrating procedural generation into your workflow allows you to generate complex initial forms, which you can later manipulate using more traditional techniques. This blended approach maximizes efficiency while enabling a high level of artistic control and precision. The key is to embrace this combination, empowering your artistry with the strengths of both traditional and algorithmic methods. The result is not only a more efficient workflow, but also a more vibrant and expressive piece of art, deeply interconnected with the sonic landscape it represents.

5.2 Sculpting Methods

Digital sculpting offers a powerful, intuitive approach to creating organic forms, perfectly suited for visualizing the fluid nature of sound. Unlike the rigid precision of polygonal modeling, sculpting allows for a more freeform, expressive workflow. Imagine shaping clay, but instead of your hands, you're using a digital stylus to manipulate a virtual medium. This immediacy is crucial for capturing the ephemeral qualities of sound – the subtle swells, the sharp peaks, the gradual decays. The process fosters a direct relationship between the artist's intention and the final form.

Many software packages offer sophisticated sculpting tools. ZBrush, for example, is renowned for its powerful brush system and dynamic subdivisions, allowing for incredibly detailed models. Mudbox, another industry standard, boasts a user-friendly interface and excellent integration with other 3D applications. Blender, a free and open-source alternative, provides a surprisingly robust set of sculpting tools, making it an accessible option for students and beginners alike. Each program offers a unique set of brushes, modifiers, and functionalities, encouraging exploration and experimentation to find the best fit for your creative process.

Understanding the underlying principles of digital sculpting is key to harnessing its potential. The process usually begins with a simple base mesh, a rudimentary shape that serves as the foundation for your sculpture. This could be a sphere, a cube, or any other primitive shape. From there, you gradually add and remove material using various sculpting brushes. These brushes simulate real-world sculpting tools, each offering unique properties that affect the surface's texture and form. Experiment with different brush strengths, sizes, and settings to achieve varied effects. Pay attention to detail; refining the smaller elements is critical for a realistic and believable result.

Mastering brush dynamics is crucial for achieving realistic results. The pressure sensitivity of your stylus translates into the intensity of the sculpting effect. A light touch might gently nudge the surface, while a firm press can carve out deep grooves. This dynamic interaction mirrors traditional sculpting techniques, fostering a natural and intuitive workflow. Experiment with the various brush options available in your software. Consider how different brushes, from smooth smoothing tools to aggressive chiseling brushes, can be used to express the subtle nuances of your audio data. This sensitivity allows the sculptor to reflect the subtleties within the audio signal, creating a visual representation that truly captures the essence of the sound.

Beyond the basic sculpting process, advanced techniques can significantly enhance your workflow. This includes using masking tools to isolate specific areas for modification and employing layers to maintain a non-destructive workflow, allowing you to revert to previous stages of your work. Projection techniques allow you to transfer details from one mesh to another or import external imagery to create intricate textures on your digital sculpture, allowing for greater precision and control over the final output. These techniques increase efficiency and enable the artist to incorporate diverse methodologies and approaches into their workflow.

Finally, remember the importance of iterative refinement. Digital sculpting is rarely a linear process; it involves constant experimentation, adjustment, and refinement. Don't be afraid to experiment, make mistakes, and learn from them. The fluidity of digital sculpting allows for continuous adjustments and experimentation, providing ample opportunities for iterative refinements. Regularly saving different versions of your work is essential in this process. The end result is a seamless blend of artistic vision and technical skill, transforming sonic information into a compelling three-dimensional narrative. This interplay between experimentation and refinement results in a dynamic and highly expressive work of art.

5.3 Boolean Operations

Boolean operations are powerful tools for sculpting complex 3D models from simpler shapes, offering a level of control and efficiency unmatched by manual methods. Imagine needing to create a complex spaceship hull – instead of meticulously modeling every curve and panel, you can combine and subtract pre-made shapes, drastically reducing workload and increasing design flexibility. This chapter will delve into these fundamental processes, empowering you to craft intricate forms with ease.

The core Boolean operations are Union, Subtraction, and Intersection. Union combines two or more objects into a single, unified shape. Picture two separate cubes; a Union operation merges them, creating one larger, potentially irregular shape. The result is a seamlessly integrated object, eliminating the need for manual merging and smoothing. This simplification streamlines the design process, saving time and resources, which are particularly beneficial for complex projects requiring multiple iterations.

Subtraction, conversely, removes one object's volume from another. Imagine carving a hole in a solid block; subtraction emulates this process, creating precise cutouts and recesses within existing models. This is invaluable for intricate details like window frames on a building, or for creating complex internal structures within a character model. Mastering subtraction allows for precise control over shape and form, adding detail without adding to the overall polygon count unnecessarily.

Intersection reveals only the overlapping volume between two or more objects. This operation creates new shapes from the shared space where objects intersect. Think of two intersecting cylinders; the intersection operation leaves only the portion where they overlap, effectively creating a new, unique form. Its applications extend to designing complex joints and intersections within mechanical models or generating organic shapes with irregular boundaries.

The effectiveness of Boolean operations is significantly impacted by the quality of your original models. Overlapping faces or non-manifold geometry can lead to unpredictable and often unusable results. Ensure your source meshes are clean, watertight, and consistently oriented. Remember that the more complex your source models, the longer the Boolean operation will take to compute. Experimentation and efficient workflow are key to mastering this technique.

Beyond the core operations, most 3D modeling software offers refinements and options. These can include settings to control the tolerance for overlapping faces, allowing for small discrepancies during the operation. Some software might provide tools to automatically fix minor issues that can arise during the process, resulting in a cleaner and more efficiently-worked model. Always explore your software's specific options and features to further refine your Boolean workflows.

Effective application of Boolean operations requires strategic planning. Don't just randomly combine shapes; consider the desired final form and plan the sequence of operations accordingly. Experimentation is encouraged—try different combinations and sequences to achieve your target model. Remember the order of operations matters; performing Union then Subtraction will yield a different result than Subtraction then Union. Mastering this sequential thinking is pivotal to efficient 3D modeling.

Let's consider a practical example: creating a futuristic cityscape. Start by creating simple shapes – cubes for buildings, cylinders for towers. Using Union, combine these base shapes to create larger structures. Then, use Subtraction to create windows, balconies, and architectural features. Finally, Intersection could be used to model intersecting bridges or complex roof structures. By strategically combining and manipulating basic shapes, one can construct a sophisticated and intricate model.

Boolean operations are a crucial element in streamlining complex modeling projects. They are not simply shortcuts, but powerful tools capable of producing results impossible or extremely difficult to achieve through traditional modeling techniques. Practice consistently to develop an intuitive grasp of these operations, improving not only your technical skills, but also your overall design process. Their impact extends from efficient workflow to the very creation of forms, leading to richer and more imaginative artistic expression. As you become more familiar with this powerful skill, you will discover an expanded range of creative possibilities. You will be able to create shapes you would never have been able to imagine without the aid of Boolean logic, truly enabling a new level of artistic expression through 3D modeling.

6 Interactive Installations

Real-time rendering is paramount for interactive installations. Imagine a sculpture reacting instantly to the music, its form shifting and glowing in sync with the beat—that's the power of real-time rendering. Engines like Unity or Unreal Engine provide the framework; you'll need to learn to optimize your scenes for smooth performance. High polygon counts and complex shaders can quickly overload even powerful machines, leading to frustrating lag. Therefore, careful planning and efficient modeling techniques are crucial. Consider using level of detail (LOD) systems, where simpler versions of your model are displayed at a distance, improving frame rates without sacrificing visual fidelity close-up. Experiment with different rendering pipelines, perhaps optimizing for mobile devices if your installation might be used in less powerful environments. Remember that audience experience comes before perfect realism, sometimes a stylized approach works better. Choosing the right real-time engine will depend on your project's scale and your familiarity with different software.

Integrating interactive elements is where your creative vision truly takes shape. Think beyond simple buttons and sliders; explore motion tracking, proximity sensors, and even brain-computer interfaces for truly innovative control schemes. A responsive installation might alter its appearance based on the audience's proximity, adjusting colors or animations as people move around it. Or perhaps viewers' gestures could control aspects of the visuals or audio. The possibilities are vast and limited only by your imagination and technical capabilities. Don't be afraid to experiment and combine different sensors; even something as simple as a pressure sensor under a platform could translate into a fascinating visual effect. You need to consider the flow of interaction; how intuitive is the experience? Will the user quickly understand how to interact with the piece? User testing is essential at this stage.

Software integration is more than just selecting the right 3D modeling program. Your project will likely require a variety of software to manage audio, visuals, and interactivity. You might use Max/MSP for audio processing, Blender for 3D modeling, and Unity or Unreal Engine for the interactive environment. Learning to work with APIs and data exchange formats like OSC (Open Sound Control) is crucial for seamless communication between these different programs. This part of the process requires a strong understanding of data flow and how different pieces of software communicate with each other. A well-designed data pipeline is essential; otherwise, your installation risks becoming a collection of disparate parts instead of a cohesive whole. Choosing software involves practical considerations. Some programs are more user-friendly than others. Consider your strengths and any available resources before committing to specific software.

Designing user interaction is less about aesthetics and more about user experience. Your installation should be intuitive and engaging. This means careful consideration of how the user interacts with the artwork and the feedback they receive. The more seamless and natural the interaction, the more impactful your installation will be. Avoid overly complex control schemes; simple, intuitive interfaces allow viewers to fully engage with the experience rather than getting bogged down in technicalities. The success of an interactive installation is largely determined by how user-friendly it is. Consider incorporating visual cues and feedback to guide the viewer's interaction with the installation, providing a clear indication of what actions they can perform. Don't underestimate the power of simplicity.

Deployment strategies depend heavily on the context and scale of your installation. A small-scale piece may be easily transported and set up, whereas a large-scale project requires more complex logistical planning. Ensure the location has adequate power and network connectivity. Test your installation thoroughly in the final environment to account for unexpected variables such as lighting, acoustics, and even temperature fluctuations. Consider issues of durability and security if your installation will be in a public space. A successful installation isn't just about the technology, it's about providing a complete and memorable experience for your audience. Think about the environment the installation will occupy, and how the installation will integrate with the surroundings. The presentation and space are as important as the technology itself. Document your process thoroughly; this will be invaluable if you need to repair or modify your installation later, or if others need to recreate your work.

6.1 Real-time Rendering

Real-time rendering is crucial for interactive installations, allowing for immediate visual feedback to your audio input. Imagine sculpting a virtual landscape that morphs and shifts in response to the rhythm of a piece of music—that's the power of real-time rendering. It's about creating a dynamic, responsive experience, bridging the gap between the sonic and the visual.

This responsiveness is key. A static 3D model, however meticulously crafted, lacks the engagement of a system that reacts instantaneously to changes in audio. Think of it like the difference between watching a film and participating in a live performance: real-time rendering makes your installation a performance, a collaboration between sound and form. The visual becomes a live instrument, shaped and molded by the very audio it responds to.

Choosing the right software is paramount. Many options exist, each with its strengths and weaknesses. Consider factors like ease of use, rendering capabilities, and compatibility with your chosen audio software. Some packages are specifically designed for real-time applications, offering optimized tools and workflows. Others might require more technical expertise or custom scripting to achieve the desired level of responsiveness. Researching the features and limitations of various real-time rendering engines will help you make an informed decision that aligns with your skills and project goals. Don't be afraid to experiment and explore multiple options until you find the perfect fit for your creative vision.

Efficient workflow is essential when working with real-time rendering. Pre-optimization of your 3D models, textures, and shaders can drastically improve performance. Complex geometries and high-resolution textures can strain even the most powerful systems, leading to lag and a compromised user experience. Experiment with level of detail (LOD) techniques, reducing polygon count at further distances to maintain acceptable frame rates. This balance between visual fidelity and performance is a crucial consideration for any real-time application. Optimize your textures; smaller file sizes will vastly improve performance without significant visual loss in many cases.

Shader programming can significantly enhance the visual impact and responsiveness of your installation. Shaders are small programs that run on your graphics card, determining how light interacts with surfaces in your scene. By writing custom shaders, you can create unique visual effects that react directly to your audio input. For example, you might use audio data to control the intensity of a glow effect, or map frequency data onto the color of a surface. This offers an incredible amount of creative control, allowing you to translate subtle changes in sound into compelling visual transformations. However, this can increase your development time, and requires a degree of programming knowledge beyond basic 3D modeling.

Hardware is another important factor. Real-time rendering is computationally demanding, requiring significant processing power from both your CPU and GPU. Investing in a system with a powerful graphics card is essential for achieving smooth frame rates. The specific hardware requirements will vary depending on the complexity of your installation and the rendering engine you've chosen. High-end processors are also important, allowing for faster processing of audio data and efficient execution of shaders. Keep your cooling system in mind; real-time rendering puts heavy loads on components that produce substantial amounts of heat.

Finally, remember that your installation's success hinges on more than just the technical aspects. User interaction design is crucial. Ensure your interface is intuitive and user-friendly, allowing visitors to engage with your work without unnecessary frustration. Consider adding interactive elements that allow visitors to influence the visual output in real time, further enhancing their sense of involvement. Think carefully about the overall experience you want to create. How can your interactive components enhance the artistic message conveyed by your installation? How can the interaction be made intuitive for the end-user? The seamless integration of interactive elements will dramatically increase the impact and enjoyment of your installation.

6.2 Interactive Elements

Designing engaging interactive elements hinges on understanding your audience and their expectations. Consider the overall experience you want to create: Is it meant to be playful, informative, or thought-provoking? A child's interactive exhibit needs far simpler controls than a sophisticated art installation, demanding a careful consideration of the user interface—the gateway between the user and the experience. For beginners, starting with a clear, concise design is paramount, avoiding unnecessary complexities that might overwhelm and frustrate. Keep it clean and intuitive, focusing on seamless interaction.

Effective interactive elements often stem from a clear understanding of the underlying audio-visual data. For instance, the amplitude of a sound could directly control the scale of a 3D object, creating a visual representation that's both intuitive and engaging. Or, perhaps frequency variations manipulate color palettes, dynamically transforming the visual aesthetic in real-time. Think of these parameters – amplitude, frequency, even harmonic content – as your building blocks. They are the raw materials of the interactive experience, providing the data flow which will drive and shape your visual response.

The choice of software greatly impacts the ease and effectiveness of integrating your interactive components. Some platforms offer built-in tools for real-time audio analysis and visual feedback, simplifying the development process. Others might require more custom scripting or integration with external libraries. It's crucial to select software which aligns with your skillset and project scope. A thorough exploration of available software options, from user-friendly environments to more complex professional tools, will save considerable time and energy during the development process. Start with simpler options, master the basics, and gradually upgrade to more advanced tools as needed.

User interaction design often gets overlooked, yet it's a critical aspect of effective interactive installations. How will users interact with your creation? Will they use physical controllers, touchscreens, motion sensors, or a combination? This decision influences not only the programming aspects but also the physical design of the installation itself. Remember that intuitive design leads to a satisfying and enjoyable experience; poorly designed interactions frustrate and distract users, hindering their appreciation of the art itself.

Deployment significantly impacts how your interactive installation will ultimately be experienced. Consider the physical environment: A large-scale public installation will require robust hardware and careful consideration of lighting, acoustics, and potential environmental factors. Conversely, a smaller-scale exhibit or presentation might have different constraints, allowing greater freedom of design. Think about factors like power sources, network connectivity, and user safety. Thorough planning at this stage prevents unexpected issues during implementation. Even the smallest detail can dramatically affect the final presentation.

While programming experience is advantageous, many tools offer visual scripting environments, which allow for interaction design without extensive coding knowledge. Beginners can start with these simpler, more intuitive tools. Learning to manipulate parameters visually through these environments allows creators to understand the principles of interactive feedback without being bogged down by complex coding. As proficiency grows, the transition to more sophisticated programming languages is then more manageable and meaningful.

Real-time rendering is essential for creating a smooth and responsive interactive experience. Stuttering or delays can severely hinder the user's engagement, creating a disruptive experience that undermines your artistic vision. Optimization techniques, such as level of detail adjustments and efficient shader programming, are crucial for ensuring smooth performance. Understanding these technical aspects is crucial, even for beginners, as the visual experience is directly affected by how smoothly the visuals update and respond to the user interaction.

The integration of various software packages often involves navigating data exchange protocols and compatibility issues. Therefore, understanding your chosen software's capabilities and limitations is essential before embarking on complex integrations. Using open-source software can facilitate collaboration and provide broader compatibility, while commercial options can offer advanced tools, streamlining the development process. It all boils down to balance and sensible project management.

Beyond the technical aspects, consider the narrative you wish to convey. Interactive elements should not simply add movement; they should enhance the story or emotional impact of your creation. Think carefully about how each interactive component contributes to the overall experience, making sure the interactivity serves a higher purpose. Always reflect on the relationship between your interactive elements and the narrative flow, ensuring a harmonious balance between technology and artistry. Your interactive element should elevate your artistic vision, not obfuscate it.

Finally, remember that iteration is key. Beginners often underestimate the importance of testing and refinement. Create a prototype, test it with users, gather feedback, and iterate on your design based on that feedback. It's a continuous process of learning and improvement. Don't be afraid to experiment, make mistakes, and learn from them. The iterative nature of design allows for refinement and improvement, crucial for creating a highly polished and engaging interactive experience.

6.3 Software Integration

Successfully integrating your audio-reactive 3D models into a cohesive, functional installation requires careful consideration of several key software choices and workflows. The right software suite can streamline your process, allowing you to focus on artistic expression rather than technical hurdles. Think of it as assembling a finely tuned orchestra; each instrument (software) plays its part to create a harmonious whole. Choosing the right 3D modeling software is paramount. Blender, a free and open-source option, offers a powerful and versatile suite of tools, perfect for beginners exploring the creative possibilities of 3D modeling and animation. Alternatively, industry-standard options like Maya or 3ds Max provide extensive features, though they come with a steeper learning curve and often a significant price tag. The decision ultimately rests on your budget, existing skillset, and the specific demands of your project. Your project's scale and complexity will largely influence this decision.

Next, you'll need software capable of handling real-time audio processing and interaction. Max/MSP or Pure Data (Pd) are strong contenders for their flexibility and power in manipulating audio signals. These programs excel at creating interactive audio elements, perfect for triggering animations or altering visuals based on audio input. Consider the strengths of each – Max/MSP boasts a more visual programming interface, while Pd prioritizes a text-based approach, offering more control over granular audio elements. Your familiarity with visual versus text-based programming languages will greatly influence this decision.

The bridge between your 3D modeling environment and your audio processing software is crucial. This is where a robust scripting language like Python or a visual scripting system within your 3D software becomes invaluable. Python's widespread use across different platforms offers unparalleled flexibility, allowing for seamless data exchange between disparate programs. This cross-platform communication facilitates dynamic updates to your 3D model in response to audio events, creating a truly interactive and responsive system. The choice depends on your programming skills and the complexity of the desired interaction.

Efficient data transfer is paramount for real-time performance. For instance, implementing OSC (Open Sound Control) messaging for communication between your 3D modeling and audio processing software minimizes latency and ensures smooth, responsive interactions. Understanding OSC's protocol and how to implement it within your chosen software is essential for creating seamless transitions between sound and visual elements. Learning to structure your data flow effectively is critical; this stage demands precision and clarity in organizing and transporting your data.

Finally, presentation is key. Integrating your installation into a user-friendly environment requires thoughtful user interface (UI) design. Software like TouchDesigner, widely praised for its real-time visual capabilities, provides an excellent platform for creating intuitive controls and visualizations. You might even consider integrating physical interfaces like sensors or controllers to further enrich the interaction and customize the user experience. Careful consideration of UI aspects is essential for maximizing audience engagement and providing them with clear pathways for interactivity. The design should be intuitive, uncluttered, and aesthetically in line with the overall visual experience of your installation. The user interface represents the direct engagement the viewers have with your creation, so designing it with a keen sense of user experience is as important as the aesthetic and computational aspects of the installation.

6.4 User Interaction Design

Designing intuitive and engaging user interfaces is crucial for interactive installations. Think of your installation not just as a visual spectacle, but as a conversation between your artwork and the viewer. A poorly designed interface can disrupt this conversation, frustrating the user and undermining the artistic impact. Therefore, prioritizing user experience is paramount to the success of any interactive project.

The first step is to carefully consider your target audience. Who are you designing for? What are their technical skills and expectations? Understanding your audience helps inform design choices. For instance, a piece intended for a museum setting will require a different approach compared to one aimed at a technology festival with a more tech-savvy crowd. This understanding is crucial for setting appropriate expectations and avoiding complexities that would bewilder less-technical audiences. Always remember that accessibility is key. A user-friendly interface should be intuitive and easy to use regardless of age, technological proficiency or potential physical limitations.

Effective navigation is another critical aspect. Cluttered or confusing interfaces can lead to disorientation and frustration. A well-structured interface guides users seamlessly through the experience. Clear labels, intuitive controls, and a logical flow are essential. Consider implementing visual cues like arrows, highlights, and visual feedback to direct the user. This is vital for creating an immersive experience that does not alienate anyone by its complexity. Avoid unnecessary complications or steps; streamlining the interaction flow increases user satisfaction. Think visually, too: use clean, uncluttered layouts, intuitive icons and an overall aesthetic consistent with your artistic vision.

Interactive elements should be responsive and provide clear feedback. Users need to understand the impact of their actions. Visual cues, sounds, and haptic feedback are all valuable tools to provide this. For instance, a subtle change in color, a brief animation, or a short sound can signify a successful action. The use of visual and audio feedback creates a more dynamic and interactive experience. A key element of an efficient interactive experience lies in immediate response, ensuring users understand the consequences of their interactions without significant delays. This reinforces the feeling of active participation and control.

Finally, remember to thoroughly test your interface. User testing is invaluable for identifying potential issues and refining the design. Observe users interacting with the installation and gather feedback. This process allows for iterative improvement based on real-world use. It allows for improvements and refinements, even with the inclusion of more subtle changes. Gathering feedback from diverse users is key to achieving a design that is both usable and enjoyable for the widest possible range of people. It is a crucial step, enabling the creation of a user-friendly experience. This constant cycle of testing, feedback, and iteration is essential for success.

6.5 Deployment Strategies

Consider your target audience. Are you presenting to a museum, a gallery, or a festival? Each venue demands a different approach. A museum setting might require a more robust, less easily damaged installation, perhaps relying on pre-rendered animations rather than real-time processing to ensure consistent performance. Galleries, on the other hand, may favor a more polished, aesthetically integrated piece, with a focus on clean lines and a visually cohesive experience. Festivals, with their often temporary and less controlled environments, require a durable, easily transportable design that can withstand the rigors of repeated setup and takedown. Adaptability is key; your installation needs to be scalable to different spaces.

Next, pre-visualization is paramount. Don't underestimate the value of thoroughly testing your installation in various environments before the actual deployment. Simulate potential issues like lighting conditions, ambient noise, and even unexpected audience interaction. Software like Unity or Unreal Engine allows for a very realistic simulation of your finished project, ensuring your work will look and function correctly. Using virtual reality technology allows for deeper immersion and feedback gathering. Take advantage of this; it can save you from numerous headaches later. Detailed planning, including potential points of failure, is your best friend. Identify those and find alternative approaches to mitigate those potential issues.

Deployment itself presents logistical challenges. Consider the physical space, its power supply, internet connectivity (if required for your installation), and even the physical movement of the components. Heavy equipment requires careful planning for transportation and setup. If you have fragile components, you need to account for that, and create a plan to deal with those potential issues. Transporting this kind of equipment requires specific containers and professional transportation services. You also need to arrange for set up and dismantle by qualified personnel. Proper documentation of every step is essential for future reference and any potential troubleshooting.

Once the installation is in place, consider audience interaction. Will your piece be hands-on, or purely observational? If interactive, clear instructions and possibly even some form of tutorial should be considered. The goal is a positive and engaging experience for all visitors. Safety is paramount; any moving parts should be secured, and clear warnings should be provided where necessary. This includes safety protocols for the installation itself, and for the surrounding space. Ensure that the entire system is grounded correctly to prevent electrical hazards. You also need to train the people who will be responsible for operating and maintaining the installation.

Finally, post-deployment review is crucial. Gather feedback from viewers, note any technical glitches, and make a plan for improvements or modifications for future deployments. Analyze the data you collect during the installation's run; this information will be invaluable for subsequent projects. Document all changes made after the initial deployment. This iterative process of refinement ensures that your installations become increasingly polished and more effective. This can also include creating a detailed report about the entire process. This may include an outline of the technical difficulties encountered and the solutions developed. This will greatly improve your future work. This is an ongoing process of learning. Each project informs the next, leading to ever more sophisticated and compelling artistic expressions.

7 Conclusion & Future Directions

The journey through "Polygonal Polyphony" has equipped you with a powerful toolkit—a fusion of 3D modeling prowess and an intuitive grasp of audio-visual synergy. You've learned to sculpt sound into form, to animate the invisible, and to translate auditory data into compelling visual narratives. This isn't just about mastering software; it's about cultivating a new artistic language.

Remember the initial hurdle of understanding 3D audio? Now, you can not only conceptualize spatial sound but also design and implement it seamlessly within your 3D environments. Your ability to manipulate soundscapes—to layer them, shape their directionality, and weave them into the fabric of your models—has dramatically expanded. The seemingly abstract concepts of amplitude and frequency now translate into tangible visual elements, a testament to the power of visualized sound. This book hasn't just provided instructions; it has fostered a creative process. You've moved from basic shapes to complex, procedurally generated worlds; from static forms to dynamic, audio-reactive animations. The potential applications are vast and extend far beyond the confines of this text. The knowledge you've gained enables you to visualize musical scores in three dimensions, to transform abstract soundscapes into immersive virtual landscapes, and to create interactive installations that blur the lines between the virtual and the physical.

Consider the possibilities: Imagine crafting an interactive art installation where a visitor's movement triggers changes in both the sonic environment and the accompanying visual display. Visualize a dynamic music video where the visuals are directly driven by the music, every note, every beat, every crescendo shaping and influencing the animated world on screen. Perhaps even design a unique educational experience where students interact with complex scientific data through immersive 3D models and audio feedback. These are just glimpses into the boundless creative expanse now available to you.

The techniques explored here—procedural modeling, Boolean operations, real-time rendering—provide the framework for complex and innovative projects. However, remember that technology is merely a tool. True artistic expression stems from your creativity, your vision, and your willingness to experiment. Don't be afraid to deviate from the established paths, to push the boundaries of what's possible, and to integrate your own personal style and sensibilities into this novel form of artistic expression.

The future of this field is brimming with exciting potential. As technology evolves, expect even more sophisticated tools, improved software capabilities, and innovative methods for integrating audio and visuals in 3D modeling. The continued development of real-time rendering engines will further enhance the possibilities for interactive installations and immersive experiences. Research into advanced audio processing techniques promises to provide new dimensions for sound visualization and manipulation. And the exploration of emerging fields like haptics and virtual reality will offer unprecedented opportunities for audience engagement and sensory immersion.

Your journey doesn't end here. This book is merely a launching pad for your creative exploration. Embrace continuous learning; seek out new software, experiment with different techniques, and, above all, let your creativity flow. The world of polygonal polyphony awaits, ripe with opportunities to transform sound into captivating three-dimensional visual experiences. The possibilities are endless; the only limit is your imagination. Explore, innovate, and create. Your artistic voice is ready to be heard. Embrace the future.

7.1 Creative Applications

Imagine a concert hall, not built of brick and mortar, but sculpted from sound itself. The undulating walls, formed by visualized frequencies, pulse with the rhythm of the music. This is the potential of marrying

3D modeling with audio; a fusion where sonic landscapes become tangible environments. We've covered the tools, the techniques—now let's explore the boundless creative applications.

Consider architectural design. Instead of relying solely on static blueprints, architects could use audio-reactive 3D models to simulate the acoustics of a space before it's built. Imagine tweaking the shape of a concert hall based on how virtual sound waves behave within its digital twin. This allows for real-time adjustments, refining the design for optimal reverberation and clarity long before construction begins. This translates into tangible cost savings and more efficient workflows, a key advantage for large-scale projects. Moreover, it enables architects to better understand how spatial audio influences the experience of the end-user; providing valuable insights into architectural design with direct feedback.

The possibilities extend far beyond architecture. Think of interactive music videos. The visuals could evolve organically in response to the music, creating a truly immersive experience that's unique every single time. Imagine a 3D landscape that transforms, mountains rising and falling to the bass line, colors shifting with the melody, creating an engaging performance that blends the visual and the aural into a singular experience. This goes far beyond standard music video production techniques, opening new avenues for creative expression in this already dynamic field. Artists can use these methods to directly translate the emotions inherent in the music into stunning, evolving visuals.

Furthermore, we can consider the realm of virtual reality and augmented reality. Imagine crafting interactive VR experiences where users can explore sound-sculpted worlds, manipulate sonic environments by interacting with 3D models, and even compose their music within a fully three-dimensional interface. This type of creation enables an unprecedented level of artist control, allowing them to engage with their work at a deeper and more immersive level. The ability to shape the environment with sound and have the environment react and respond to sound opens up a wealth of new narrative possibilities and opportunities for user interaction, thus broadening the scope of VR and AR applications.

Education provides another fertile ground for these techniques. Students could visualize complex scientific data—say, seismic waves or atmospheric pressure—as 3D models reacting to audio recordings of natural phenomena. Instead of simply seeing numerical data on a screen, they could experience those concepts directly, enhancing their understanding and making learning a far more engaging process. This method offers a tactile and multi-sensory approach to knowledge acquisition that's more conducive to learning than traditional methods. Instructors could create custom visualizations and simulations specific to their curriculum, tailoring the learning experience to each individual student.

The applications extend to the arts, too. Imagine a sculptor whose work responds to the sounds of their environment. Or a painter whose brushstrokes are guided by the harmonies and rhythms of a musical piece. These are not just hypothetical scenarios; they are within reach, opening a myriad of possibilities to create unique art pieces using the techniques explored in this book. This opens up exciting opportunities for artists to explore new forms of expression and push the boundaries of what art can be, fostering cross-disciplinary creativity.

Finally, consider the potential for creating interactive installations for museums or public spaces. These immersive environments could combine 3D models, audio-reactive animations, and responsive interfaces to create unique and unforgettable experiences. Imagine a space that transforms based on the presence and actions of visitors, creating a personalized and evolving experience for each attendee. This form of interactive art engages audiences on multiple levels, providing a unique form of entertainment and educational experience that's both engaging and stimulating. The potential for these installations to create engaging and meaningful artistic experiences for the public is immense. It is the future of interactive art installations.

7.2 Further Exploration

Imagine a soundscape, not just heard, but seen. This isn't science fiction; it's the power you've unlocked through this book. We've built the foundation, explored the tools, and mastered the techniques. Now, let's push further. Consider your project's potential for expansion. Could your existing audio-reactive animation be enhanced by incorporating user interaction? Perhaps a simple button press could trigger a cascading effect, visually representing a sonic crescendo. Or, imagine a more intricate system, where a user's movement within a defined space alters both the soundscape and the corresponding 3D model. The possibilities are vast, limited only by your imagination and technical skill.

Moving beyond simple interactions, consider integrating external data sources. What if your visuals responded not just to pre-recorded audio, but to live data streams? Think of real-time stock market fluctuations shaping a mountain range's height or environmental sensors influencing the color and texture of a virtual forest. This leap requires a deeper understanding of programming and data manipulation, but the artistic rewards are immense. Research libraries such as Processing and openFrameworks provide valuable frameworks for this exploration. This isn't about simply creating pretty visuals; this is about telling stories with data, weaving intricate narratives through the interplay of sound and sight.

Let's delve into the realm of procedural generation. Remember the procedural modeling techniques introduced earlier? Expand on that foundation by developing algorithms that generate complex 3D models based on musical parameters. A simple melody could translate into a sprawling cityscape, each note dictating the height and density of buildings. A more complex composition might shape an entire landscape, its rhythms shaping the terrain and its harmonies influencing the coloration of the environment. The ability to generate infinitely diverse models from a single musical source opens up boundless creative potential. The key here is to start small, experimenting with simple relationships between audio data and geometric forms before tackling more complex generative systems.

Collaboration is crucial in pushing creative boundaries. Consider teaming up with other artists, perhaps musicians, dancers, or other visual artists. This interdisciplinary approach can lead to unexpected synergies and innovative outputs. The musician's composition could drive your 3D model's animation, while a dancer's movements could, in turn, influence the sound design. This collaborative process not only fosters creative growth but also allows for a richer, more layered expression of ideas. Explore the possibilities of collaborative installations, combining your expertise in 3D modeling with the creativity of others.

Finally, consider the broader impact of your work. Your artistic creations are not confined to the digital realm. Explore ways to translate your skills into physical installations and interactive experiences. Imagine a museum exhibit where visitors can interact with a 3D model that responds to their movements and sounds. Or perhaps a large-scale public art installation, where sound and visuals intertwine to create an immersive, shared experience. Think beyond the screen; consider the possibilities of augmented reality and virtual reality to further expand your reach and immerse your audience in your unique creative vision. The limitations are not in the technology, but in the scope of your imagination and the boldness of your ideas. Don't hesitate to experiment, to fail, to learn, and to redefine what's possible. The journey of artistic creation is ongoing, and your potential is boundless.

www.ingramcontent.com/pod-product-compliance
Lightning Source LLC
LaVergne TN
LVHW020236210926
848481LV00016B/2343
9798349230936